An AMA Management Briefing

Joint Ventures for Product Innovation

LEE ADLER
JAMES D. HLAVACEK

amacom
A Division of American Management Associations

Library of Congress Cataloging in Publication Data

Adler, Lee.
Joint ventures for product innovation.

Includes bibliographical references.
1. Joint adventures. 2. Technological innovations. 3. New Products. I. Hlavacek, James D., joint author. II. Title.
HD2741.A36 658.5'75 76-10830
ISBN 0-8144-2191-1

This Management Briefing has been distributed to all members enrolled in the Manufacturing Division of the American Management Associations. A limited supply of extra copies is available at $5.00 a copy for AMA members, $7.50 for nonmembers.

First Printing

Contents

About the Authors

Lee Adler

James D. Hlavacek

Lee Adler is currently Professor of Marketing in the College of Business at Fairleigh Dickinson University. Professor Adler has had 24 years of industry experience, including Director, Corporate Marketing Research, at R.C.A. and Director of Marketing Research in the Inter-Public Group of Companies. He is the author of two textbooks and numerous professional business publications, and he is a past vice president and director of the American Marketing Association.

James D. Hlavacek is a faculty member in the School of Business at the State University of New York at Albany and was previously employed by Hooker Chemical & Paint Corporation in market development. Dr. Hlavacek is Chairman of the Venture Management Section of the American Marketing Association, has written many articles for well-known business publications, and is a frequent speaker at executive development programs. He is the editor-in-chief of *Industrial Marketing Management* and has been a consultant to numerous corporations in the area of new business management.

Introduction

The domestic joint venture is an organizational mechanism for developing, manufacturing, and marketing new products. Joint ventures are a new organizational approach and entity in which two parent firms pool technology, manufacturing know-how, marketing skills, capital, and managerial expertise. As such, they enable new ideas and technology to be produced and commercialized that otherwise might never come to be.

With the risk level for innovation mounting, even for large firms, new-product development and the accompanying market development often slacken. This state of affairs is made particularly grave at a time when both domestic and foreign competition is intensifying. The pooling of company resources into synergistic joint ventures is an approach that offers great promise. As Peter Drucker recently wrote:

> The joint venture is the most flexible instrument for making fits out of misfits. It will become increasingly important. It is at the same time the most demanding and difficult of all tools of diversification—and the least understood.[1]

Not that joint ventures are a cure-all for a languishing product- and market-development function within one firm. Rather, they represent a new or-

[1]Peter F. Drucker, *Management: Tasks, Responsibilities and Practices* (New York: Harper & Row, 1974), p. 720.

ganizational approach or alternative when a single firm considers an internal venture, an acquisition, licensing, divesture, or abandonment.

Joint ventures present a unique opportunity for corporations of different sizes to benefit from each other's technical, production, commercial, and financial advantages.[2] For example, the small manufacturing corporation typically has rather narrow product lines, if not a single product line, as compared with the broad diversification of the larger and more complex corporation. And while it frequently has ideas or technologies that exhibit strong potential in end-use applications outside its current markets, the small firm's restricted focus is usually accompanied by restricted marketing, financial, and production capabilities that cannot effectively commercialize a product innovation outside its existing operations. Furthermore, such a firm is often technology-based and may have no real interest in developing production efficiencies and marketing capability in a new area.

The joint-venture approach allows each parent firm, regardless of its size, to participate in product innovation in the areas it knows best. It utilizes the previous investments and experiences of two firms in a symbiotic arrangement that opens avenues that might otherwise be barred to individual firms. It may thus enable new technology to be developed, increase productivity, and deploy corporate resources more efficiently.

[2]Edward B. Roberts, "Technology Strategy for the European Firm," *Industrial Marketing Management*, August 1975, p. 197.

1
Background and Need for Joint Ventures

Most budgets run shy 20 percent, except for renovation and vacation budgets, which are typically 50 percent low. In ordinary times such a procedure is adequate to meet the challenge. But, when slackening sales volume and profit, diminished cash flow, and the high cost of money are compounding the problems generated by other inflated costs and the new-product-development function is especially lagging in many companies, a budget needs to be expanded to meet the challenge.

To these financial problems can be added others, some new and some old, that constrain product innovation:

—Organizational difficulties—such as disharmony among technical, production, and marketing people, jurisdictional squabbles between operating divisions, and conflicts between staff and line groups.

—Materials shortages—petroleum and many nonpetroleum-related materials are in short supply and often are cartel-controlled with high and unstable prices; this is especially critical when the materials make up a high percentage of the value-added output.

—The lack of some essential resource—whether in production, R&D, or marketing; this is especially true when R&D is working alone too long and from the beginning has failed to consider the existing manufacturing capability to make the device economically in the quantities expected to be shipped to the marketplace. The time and cost to develop a new distribution system or "just" train the same or a new sales force to sell the product are often staggering.

—Simply running out of creative steam—often a too-close and too-long application to a particular project induces staleness and slackness; a view from outside is frequently more objective and can recharge the enterprise.

—Complete or partial bailout—recourse to divesture or licensing as a quick and relatively easy solution to product-development problems often greatly reduces the longer-term downside potential.

This state of affairs is regrettable because the need for continued infusion of new products is at least as great as it ever was. Competition does not just go away; if anything, it intensifies. Uncontrollable external factors persist—

political and legal constraints, pressures to exercise greater social responsibility, the demands of the consumer movement.

Moreover, current products inevitably age and decline in favor because of changing customer requirements, life-styles, and tastes. Indeed, much of the recent history of U.S. industry seems to reflect the accelerating erosion of entrenched positions prompted by the interlaced workings of all these forces.

Alert companies are by no means oblivious to this situation and have been taking a number of corrective steps. Some are seeking to rejuvenate current products, to find new market segments for existing products, or to promote new applications in established markets. Others, in their more sophisticated approach to utilizing the marketing concept, are evolving and selling systems, committing themselves not only to functional but to financial performance as well.

Yet other firms are reorganizing so as to attune themselves more acutely to their markets and thus become more sensitive to the needs of their customers and better able to serve those needs by marshaling their talents.[3] And still other companies have travelled the merger/acquisition route as a way of acquiring and developing innovations.

However, of all the mechanisms used in recent years to spur significant new-product development, one of the most common has been the intracompany venture team. Unfortunately, despite its theoretical allure in bringing together the required mix of disciplines in a quasi-independent group, the venture-team approach has often not worked too well in practice. Both top management and venture-team managers agree on three reasons for the frequent inadequacy of this approach:

Market too small—what constitutes a large or small market varies relative to the size of the firm producing and marketing the product. One new product with a potential of $10 million in annual sales in a $2 billion company was termed a "junior achievement product" even though R.O.I.s were projected to be better than most areas of the firm.

Distribution problems—the reluctance of existing divisions or sales forces to take on the new product, which involves different customers who further must be educated of the product's benefits; the problem of recruiting competent distributors and/or service support is slow, costly, and often haphazard in the new market in which you have little experience and no reputation.

Internal problems—product managers and divisional managers are evaluated on quarterly and yearly profit performance from which a new product is usually a cash eater; the more mature and stable product lines are often in older plants (that may have excess capacity) where resistance to newness or change is great.[4]

To these, top management adds two other major reasons: cost overruns and disappointing profits.

[3]See Mack Hanan, "Reorganize Your Company Around Its Markets," *Harvard Business Review*, November-December 1974, pp. 63ff.

[4]James D. Hlavacek, "Toward More Successful Venture Management," *Journal of Marketing*, October 1974, pp. 56–60.

Given these distressing internal experiences, a growing number of firms have recognized the potential benefits of cooperative arrangements with external partners. In doing so, however, many managements have no trouble hiding their enthusiasm for outsiders.

The explanation for this is largely psychological. Some executives suffer from the NIH ("not invented here") syndrome; the fear, often unconsciously, that the need to seek help outside the firm will be perceived as an admission of corporate or personal weakness. Other executives, more emotionally secure within themselves, may fear that their existing staffs will feel threatened.

Nonetheless, under the imperative to continue to innovate, poolings of resources between independent firms are inevitably increasing in number and scope. The many different modes of allying the resources of two firms so as to enhance the strengths of each fall into two major categories.[5]

The first is exemplified by one company dealing with another on either a one-shot or a long-term basis but without any major innovation in organization. Examples are sharing facilities such as a distribution network or physical plant, franchising, or licensing. The second mode occurs when two companies create a third entity; this may range from a temporary joint study group or tie-in advertising campaign all the way to a separately organized corporation owned by the two parents.

It is this last instrument for product innovation that is of particular interest here. To identify the major factors underlying the successful use of joint ventures for this purpose, a field study was conducted. The findings and implications of that study comprise the subject of this monograph.

The joint-venture form of organization is often called into being (1) by the oil and gas industry for the joint exploration, extraction, and/or distribution of natural resources; (2) by the real estate development industry in its approach to a major construction project requiring a large amount of resources; or (3) when a firm in one country establishes a production and/or marketing system in another country in alliance with a second film usually from the country.

But in this study we are focusing on the creation of a new organizational entity designed by two corporate partners to develop, manufacture, and market a *product* innovation new to one or both of the parent corporations.

[5]Lee Adler, "Symbiotic Marketing," *Harvard Business Review*, November-December 1966, pp. 59ff.

2 Field Study of Joint Ventures

Identifying joint ventures established to create, manufacture, and market new products is very difficult because they are privately held by the parent firms. As a result, public information relating to their existence, location, size, and activities is generally very limited. Following standard accounting procedures, the joint venture may only be footnoted in the annual report of each parent firm or "hidden" in their income statements as "other income."

Joint ventures for product innovation were identified from a lengthy search of annual reports and business periodicals and from professional and trade association contacts. Twenty-three new-product joint ventures were located over the last six years. One or both of the parents were then approached to validate the joint venture's existence, determine its geographic location, and obtain the name of the joint-venture manager.

Four of the 23 joint ventures had been defunct for one to three years and were alliances of extremely small parent firms; in two of these cases, both firms had either moved or been dissolved. This resulted in a viable sample of 19 joint ventures. Two of the 19 either had just been liquidated or were about to be dissolved. The current venture manager (or previous manager in the dissolved ventures) was contacted for a personal interview appointment.

All 19 venture managers agreed to participate in the personal interviews. The representativeness of this sample frame may be debatable, but without any better existing sample and without any previous field research in the area as a guide, these 19 were the sample selected.

In a later phase of this study a number of attorneys associated with the formation of many of the joint ventures studied were interviewed by telephone or in person to determine the legal considerations common to joint-venture formations.

Classification

Fourteen of the 19 joint ventures studied, or about 75 percent, fell into three two-digit Standard Industrial Classification (SIC) groups—paper products (SIC 26), chemical products (SIC 28), and primary metals (SIC 33). Nine (close to 50 percent) were engaged in the production and marketing of chemical products, primarily fibers, plastics, and rubber products (SIC 282), basic chemicals (SIC 281), and drugs and medicines (SIC 283).

A possible explanation for the large number of joint ventures in the chemicals industry lies in the necessity here for diverse technical abilities and special knowledge in developing new fields or opening new markets.[6] Backman found joint ventures to be even more extensive in the petrochemicals field. Both petroleum and basic chemicals companies had advanced know-how in various areas of petroleum and chemicals technology and could pool that expertise, but the chemicals companies had the advantage of an established distribution system into which the petroleum companies wanted to introduce a new diversity of products.

One of the authors[7] also found the chemicals industry to be a major user of the joint-venture approach for "symbiotic linkages" because of the high financial burden of R&D and the need to economize by adapting another firm's technological expertise or marketing know-how to its own industry. Joint ventures for new-product innovation are probably less prevalent in those industries whose survival is less dependent on a solid technological foundation.

Exhibit I on the following pages shows a number of the cases analyzed relative to each parent organization's contribution, the new product, respective parental size, and the length of the negotiation period.

[6]Jules Backman, "Joint Ventures in the Light of Recent Antitrust Developments: Joint Ventures in the Chemical Industry," *Antitrust Bulletin*, January-April 1965, p. 14.

[7]Adler, op. cit., p. 36.

Exhibit 1. Sample of domestic new-product joint ventures.

Parent Firm's Contribution	Parental Size in Annual Sales ($) When Venture Was Formed	Negotiation Period Before Formation	Joint Venture's Mission
A. Coal conversion technology	4.1 billion	13 months	To develop and offer efficient coal conversion plants for utilities with technology that was shelved 10 years ago.
B. Chemical process construction expertise	1.4 billion		
A. Polyester recovery technology	3 million	11 months	To produce and mold compounds from recycled plastic materials.
B. Scrap chemicals business	10 million		
A. Black-and-white to color film process	2 million	8 months	For black-and-white film transmitted to viewing over color television.
B. Nationwide television network affiliates	880 million		
A. Self-injection drug-cartridge technology	3 million	10 months	A self-administered heart-attack drug and injection system.
B. Nationwide pharmaceutical sales distribution system to heart specialists	1.7 billion		
A. Financial news data and nation-wide sales force to stockbrokers	170 million	11 months	An up-to-the-minute computerized financial information system for stockbrokers.
B. Computer systems for the banking and financial community	250 million		

A. Advanced welding technology	3.0 million	7 months	To market an advanced welding process that has resulted in many applications.
B. Worldwide welding products and metallurgical business	350 million		
A. Molten-salt pollution-control process	1 million	12 months	To design and construct antipollution systems.
B. Construction division with nationwide sales force in waste and water treatment	400 million		
A. High-density-polyethylene technology	3.8 billion	14 months	To produce and market synthetic plastic printing paper.
B. Integrated paper manufacturing with full market coverage	1.2 billion		
A. Batteryless and wireless call boxes	1.5 million	3 months	To produce and market emergency call boxes.
B. Highway products division with nationwide distribution and maintenance service	2.4 billion		
A. Thermoplastic photosensitive film—erasable or updatable microfilm process	100,000	15 months	To produce and market an updatable microfilm process.
B. Nationwide office products distribution-sales force with over 700 salesmen	600 million		

Parent Firm's Contribution	Parental Size in Annual Sales ($) When Venture Was Formed	Negotiation Period Before Formation	Joint Venture's Mission
A. Automatic transmission technology for heavy-duty trucks	375 million	10 months	To produce and market automatic transmissions for heavy-duty trucks.
B. Major truck-engine company with worldwide sales-service centers	605 million		
A. Centrifugal compression technology	3 million	2 months	To produce and market a high-speed machine with centrifugal compressors application.
B. Large air-conditioning and compressor company with marketing skills and production capability	800 million		
A. Petrochemicals technology	2.7 billion	12 months	To produce and market thermoplastic resins.
B. Plastic polymer production and marketing capability	433 million		
A. Computer peripheral equipment	5 million	6 months	To produce, market, and service "point-of-sale" electronic checkout systems for retail stores.
B. Nationwide distribution and service system to major retailers	301 million		

3 Study Findings

Reasons for the Joint Venture

Most of the firms cited multiple reasons for entering into a joint venture. The principal ones were:

—To acquire needed expertise in either production, technology, and/or marketing.

—A need for additional capital—especially in the case of the small technology-based firm; but if the other partner provides *only* capital it is an investment, not a joint venture. This study is not concerned with the arrangement whereby one partner contributes only capital.

—A preference for concentrating on the special skills of the firm. We typically encountered this desire on the part of the small high-technology firm that often had no real interest in developing marketing competence, preferring to spend its time and resources doing what it knew best.

—To speed up the process of new-product and market development, enter the market earlier and more intensely, and bring in an earlier and larger cash flow.

—To facilitate diversification—by placing the technology-contributing parent in a new marketplace with the parent already there.

—To reduce the risk of being challenged by the Justice Department for being in violation of the antitrust laws, as would more likely be the case if some other means were used, notably acquisition.

The major internal barriers faced by the small firms were the size of the capital requirements, the need for production know-how and capacity, and the need for a marketing system or more extensive distribution network to reach a more diverse market than was currently being served.

But even the larger firms in our sample, confronted by inflation, high interest rates, and an arid equities market, have been forced to reevaluate their growth and expansion avenues. For example, a national financial news service firm considered developing an electronic news retrieval service for the financial community but decided it would be technically too difficult and too costly to develop such a system alone and subsequently formed a joint venture with an established computer manufacturer that was well known as a specialist in computer systems for the banking and financial communities.

With stock often selling between 25 and 40 percent of book value and with high debt levels to cover, corporations have found that growth through internal product development or via acquisitions is considerably more difficult. These depressing factors have also caused a more rigorous application of the various criteria used to assess new-product potential, for example, anticipated payout period, compatibility with existing product line and marketing facilities, drain on available manpower, and so on.

As a result, more new-product concepts are being screened out or shelved. As a further result, the impetus to seek a marriage of resources with a co-venturer is intensified. Not surprisingly, then, the typical alliance in our study consisted of technology supplied by one parent and marketing and financial resources furnished by the other.

Significantly, nearly all the firms considered other approaches before seeking a joint venture. All the large firms seriously pondered independent development first, and four of them invested substantial funds over many years in unsuccessful internal attempts.

Licensing was also examined as an option—especially by the smaller firms—and was seen as a way of providing a more rapid income over the short term to help recover expenditures on costly research with no additional risk. But it was also perceived as providing a lower return on investment potential than a joint venture. As on joint-venture manager remarked, "We would've been able to achieve only about 5% of the financial return from a licensing arrangement which would've also had a shorter life than the joint venture."

The larger partner in the joint venture frequently attempted to acquire the smaller partner, or alternatively tried to purchase that part of the smaller firm it especially wanted—that is, some technology and/or production know-how.

Of the few attractive candidates not yet acquired, many are private or closely held corporations. If such a firm were to be acquired, historical experience shows that its competent management team would probably soon leave the merged unit. The reason is that large bureaucratic organizations, with their strict financial reporting systems, endless red tape, and procedural rigidity, caused the managers of the acquired firm to seek the less structured entrepreneurial environment of the small firm they founded and built in the first place.

If this team leaves the business soon after it is acquired, the larger firm may then have difficulty managing a business it knows little about. The joint venture, which is a *partial* acquisition for *both* parents, is proving to be a viable alternative when an outright acquisition is not desirable from either partner's perspective.

Exhibit 1 presents a sample of the contributions of each parent in the joint venture. In 14 of the 19 joint ventures studied, one parent mainly contributed technology and production know-how while the other mainly contributed marketing know-how and sometimes contributed capital.

The study showed, not so much that marketing firms discovered unfilled needs and then went looking for a suitable technology, but primarily that technological firms with underutilized technology went seeking another firm that might apply, commercialize, or market that technology as finished products. It was therefore the technology contributor who took the initiative—seeking primarily marketing capacity. In the few cases where the marketing contributor provided the initiative to establishing a joint venture, it was seeking new or better technology to complement or strengthen its existing product lines for an existing sales force or an established distribution system.

Developing Search Criteria

Even before beginning to contact other firms, it is crucial to think through one's own assets and liabilities so as to be able to articulate the requirements of a co-venture. Typical criteria include:

—An established marketing/distribution system in the market to be served. For example, the transmission division of an original equipment manufacturer sought an ideal truck-engine company that had an interest in automatic transmission for heavy trucks. A well-known truck engine company had an excellent reputation and a nationwide truck-engine sales-service network but did not have a transmission business.

—A sales force of suitable size, caliber, and image calling on specific customers. This was illustrated when a small medical products firm embarked on a joint venture with a large division of a major pharmaceutical corporation that was well known to heart specialists and had a sales force of nearly 1,000 detail men who were perfectly positioned to introduce a new medical device.

—Technology to improve on or complement one's own current technology base. As one joint-venture manager said, "Our commercial partner provided us with 90 years of knowledge and experience in the paper business."

—The kind of personnel needed—managerial, technical, production, and marketing.

—A given minimum available financial resource. This was nearly always a relatively larger partner that had a greater cash flow, access to vast capital, and staying power through adversity.

—Company size. Corporate size is a very relative matter. A company with annual sales of $5 million might perceive a company with yearly sales of $100 million as large because it is 20 times larger. The $100 million company might in turn call itself small relative to a $6 billion company. Therefore, "large" and "small" portray the *relative* differences in size between the two respective joint venture firms; and since the parental size factor is so critical, a more complete exposure is necessary.

Parental Sizes

Approximately one-half of the 19 joint-venture managers interviewed believed parental size had little or no effect on the outcome. However, many of the managers stated that symbiotic arrangements between two small technology-based firms with limited product lines tended to magnify their common weaknesses. Moreover, the basic complementary contributions and synergy required are difficult to achieve unless one of the partners is strong in marketing and financial resources.

The latter proved generally to be a large organization. Since nearly all the joint ventures studied were between two firms that would be considered "large" and "small" relative to each other, we might conclude that the joint-venture approach to product innovation is best suited to firms of differing sizes, with one's strengths more than complementing the other's weaknesses.

The large company with relatively narrow product lines is often more open to joint venturing because it is more in need of expanding its lines and technical base. It will thus encounter less technical risk in developing and offering a new product that also fits into its existing marketing and production systems.

Many large companies have pruned both old and new low-profit products or businesses and therefore need new products to "plug into" their existing sunk or fixed investment, for example, salesmen, distribution centers, plant, equipment, and technical support. The joint venture's new product may fill idle capacity, replace lost profits from pruned products, and keep the production and sales-service forces busy with the new item.

The larger parent firm in the joint venture usually supplies sophisticated staffs that can analyze the joint venture from technical, production, marketing, and financial aspects in a business or market they have been in, are in now, and plan to be in for a long time. This past experience results in more accurate estimates and educated forecasts that are probably conservative and represent low risk.

When two contributing firms of unequal size form a new-product joint venture, many potential problems can be envisioned. The larger firm with its greater financial resources might initially or eventually dominate the joint venture. Conversely, the smaller firm would be more entrepreneurial in its approach and might resist the more structured procedures that guide many of the larger firm's activities.

The joint-venture managers in our study were somewhat mixed in their views about the effect of parental size differences on the joint venture, but nearly all agreed that most potential conflicts from such differences could be avoided by anticipating the conflicts within the detailed joint-venture agreement. The variety of their views is best captured in a sample of their statements on the subject.

> The most important fact the small company must realize is that it is small. Being small, it is usually more hungry and needs the sales more. Since it is more dependent upon the new sales, it weakens its position by giving a portion of that dependency to a large company that is neither small nor hungry.

Largeness means bureaucracy, bureaucracy means time, small companies don't have time. Also, if the joint venture goes sour, the small company especially needs a mechanism to get out quickly.

The small company should never depend on the large company to sell its product or keep it alive. Continue to do your own production and marketing up to the level of your own normal operation. Gear up only to hard orders.

The large company, especially its technical people, saw the joint venture as an admission of inability or failure on its own part. Small companies' technical people are generally more receptive. They know they can't do everything themselves and therefore keep their eyes and ears wide open for any new development on the outside that they could possibly use.

The executives at the much smaller parent are difficult to do business with. They have no interest in or time for our detailed planning studies, except in the area of profitability.

We [the larger parent] need provisions for performance from our partner company as it also does from us.

The small company worked eight days a week to get the joint venture formed while the larger parent dragged its feet and needed many levels of review or decision.

Identifying Potential Partners

With search criteria in hand, each joint-venture manager was asked what steps were taken to find a partner. Eight of the nine largest joint-venture partners (five of which were not the initiators) performed detailed independent technical and market studies before forming the joint venture. Five of the smaller joint-venture partners stated that they did not have either the time or the personnel to conduct such formal evaluations.

Generally, the small initiating firms tended to locate a partner through more informal methods than the larger firms employed. These methods included trade shows and a board member's knowledge of another firm. Leads to help either large or small initiating firms to identify potential partners included:

—The "trade," usually meaning other industry members met at trade or professional association meetings.

—A competitor—if one partner soloed with the joint venture in the marketplace, it knew the most competent competitors.

—Current or former suppliers—a close relationship had developed from previous business transactions, resulting in each partner knowing the other.

—Board members—from their business contacts or business research, they knew of suitable partners.

—Nonexclusive licenses—the joint venture was a means to better utilize this previously limited relationship.

—Sales force observation—when the technology-contributing firm attempted early, limited marketing alone, its own or the competitor's salesmen brought the lead back to headquarters.

—Information from the *Thomas Register*—the major firms in a business area or product line are shown or listed in this directory.

—Information from a market research study, usually to confirm the *Thomas Register* source, with a relative ranking of the firm's strengths and weaknesses in this area.

Through the use of one or more of these identification procedures, the initiating firm was able to list and rank their partner preferences. It is interesting that none of the joint ventures analyzed resulted from the efforts of a broker or "professional finder". One firm initially employed a finder, but became dissatisfied with this method and eventually located more appropriate prospects through its own efforts.

With potential partners identified, nearly half of the initiating corporations were unable to romance their first choice into a marriage. This was exemplified by one firm that ultimately had to settle for its third choice—the first was too committed to other new products and the second simply stated that it was not interested. A "no" in the early courting stage can also depend on *who* was contacted in the potential-partner firm and on *who* conducted the search at the initiating firm.

Conducting the Search for a Partner

The larger corporations typically had a corporate development or planning office at the corporate level—in the very large firms, at the divisional level—to identify potential partners and conduct the search. By contrast, in the smaller firms top management carried out the search.

The execution of the search for partners among those already identified on paper appears to be very much an art. As stated previously, a "no" from a first or second choice may be a function of *who* in the potential-partner firm was contacted. In the case of one large firm, a "no" from the vice president of R&D was transformed into a "yes, we certainly are interested" from a division general manager who ultimately closed the deal to form the joint venture.

The profit-motivated and profit-responsible division manager was interested because his evaluation was based largely on financial performance, not on how he achieved the financial results. Therefore a turndown by a *person* should not be viewed necessarily as a turndown by the *company*. It took this one small technology-contributing parent firm nearly three months to locate the *key* person at the much larger parent firm. Persistence is necessary to locating the receptive person at the potential-partner firm.

It was common for the smaller technology-contributing firm to contact a high-level technical or R&D executive in the potential-partner firm. But a "no" from the large R&D function was often motivated by the perceived problems a smaller firm would bring to the company's laboratories—the old NIH factor. The marketing director or appropriate product manager in the larger firm thus appears to be as good as the division manager to go to for acceptance during the courting period.

When the small technology-based firm negotiates with the large company at a lower operational level, the parties in the discussions must consider what will "sell" the scheme to superior level(s) in the large firm. It may therefore be best to first contact the president of the large company and so avoid having to work up through a number of levels in the firm.

The president of a large firm also has more latitude in taking risks than lower-level executives. He is involved in major investment decisions and often thinks in longer time frames than his subordinates. During negotiations, he can put one in contact with the most appropriate division manager, marketing manager, or liaison people. Directives from the president or CEO's office usually receive prompt attention and thorough consideration.

Negotiating the Joint Venture

In nearly all the cases examined, the initial meeting and all later negotiations were conducted by high-level executives. One joint-venture manager with success in a previous joint venture remarked: "The basic negotiations must be carried out by people who have the authority to negotiate or else time and interest are lost." In large multidivision corporations it was the general manager of the division with sunk technology-development costs on its books who initiated the search and negotiations. The basic elements to be considered and agreed upon invariably included these:

—Technical and/or commercial contributions from each partner.
—Initial capital contributions and further financial obligations.
—Key personnel desired.
—Growth and profit objectives.
—Cost accounting, pricing, and profit-and-loss distributions.
—Type of decisions requiring board approval.
—How to measure the venture's accomplishments and review the performance of the key personnel.

When two firms are negotiating to form a new firm for the mutual benefit of both, a full and early disclosure of all technical information is necessary before the joint venture is formalized. Eventual success requires basic agreement as to what the existing state of technology can and cannot do. One joint venture in our study was a disaster and another is still struggling to survive because full disclosure of the technology was not made before the new enterprise was formed.

In a third case, a firm oversold the merits of its technology to its joint-venture partner and the additional increments of development at a significantly greater cost nearly caused the parties to dissolve the relationship. In yet another case, the marketing parent of the joint venture was unaware of the premature stage of the other firm's technical contribution.

Co-entrepreneurs must exercise patience. The length of time between the initial conception and the formal birth of the joint venture (the negotiation period) ranged from one month to almost two years, with an average gestation period of ten months (see Exhibit 1).

During this courting stage, it was common among the large firms to perform independent detailed technical and market studies. These studies provided valuable information for both partners to share in the final stages of negotiations. The younger of the ventures we analyzed sometimes went through a joint-development period of one and one-half to three years before the new entity was formally created.

Because of the more thorough search for and screening of potential joint-venture partners and the slower reaction time found to be habitual in the large corporations, it is safe to generalize that as the individual size of each partner in the joint venture increases, the length of time from initial contact to the completed joint venture also increases. When a small firm joint-ventured with a much larger firm, the period to formal closing was less than when two larger firms did so. Firms that had previously entered into joint ventures were able to have their formal closing earlier; but our study suggests that the longer courting period, trial marriage, or "shacking-up period" adds to the chances of a more successful formal marriage.

4 The Joint-Venture Agreement

The legal document emanating from the negotiations plays a critical role in the downstream success of the joint venture. First, and probably most important in a joint-venture agreement, a detailed and precise definition of the joint venture's purpose or mission must be mutually agreed upon. This will help keep the joint venture focused and minimize potential conflicts of interest between the parent firms.

Any activity outside the previously defined mission should be eliminated. If it was an overlooked area, that portion of the joint-venture agreement (and any other affecting areas not covered by the initial agreement) should be rewritten.

It is crucial to write a closed-end, well-defined charter for the new enterprise, spelling out each parent's contributions, operational procedures, authority, and objectives. Many joint-venture managers and their legal counselors also stressed the need for precise specifications on how profits will be distributed and—equally important—how failure will be dealt with. A sound joint-venture agreement furthermore provides guidelines as to whether, how, and over what time period one of the partners may increase or decrease his participation in the venture.

Contributions

Our in-depth interviews revealed that the parent firm possessing the technology, regardless of the relative size of the other parent, had the greater leverage in negotiating its equity-percentage position. Organizations strong **in the marketing arena are relatively easy to find, but the firm with the technical contribution has a rather scarce asset. Possession of this asset frequently placed the small company on an equal or better-than-equal negotiating basis with a major corporation.**

Six of the 12 attorneys interviewed stated that the small firm bringing the major technical contribution to the joint venture often believed it was entitled to a majority control position. But the large firm was cognizant that the small firm frequently could not fully commercialize the technology by itself, or that if it attempted to it would do so at a much slower rate and with a greater risk than if it were a shared experience.

In the joint ventures between a large and a small firm, the technical contribution nearly always was provided by the small firm for equity, whereas the large firm provided its marketing expertise in addition to an up-front capital contribution for its equity sufficient to match the original sunk technology investment cost of the smaller firm.

Changes in Involvement—Equity Positions

Approximately one-half (10) of the 19 enterprises studied began with each of the two parent firms having a 50 percent equity position in the joint venture. The others began with percentage agreements of 80-20, 60-40, 51-49, 33-67, and so on. A number of legal clauses in the initial joint-venture agreement provide more insight into the equity interest position than the surface 50-50 or 60-40 ownership may suggest.

For example, upon formation of a 50-50 joint venture, Company A paid Company B (the smaller firm with the technology) $1,750,000; this was nonrefundable and unrelated to any future activity of the joint venture and was reported by Company B as income. The net profits of the joint venture are to be divided 60 percent to Company B and 40 percent to Company A on the first $5 million; thereafter, net profits are to be divided equally. In another example, a 60-40 joint venture, any net loss in any year is to be borne by the larger firm with the 40 percent equity position.

After a period of from one to six years following their formation, five of the ten joint ventures with initial 50-50 agreements no longer remain in that position. In two cases, one of the partners later realized it could not continue to provide financial contributions or investments on an equal basis. In another, one partner subsequently concluded it did not really want to spend more money on the enterprise because the new business area was so far removed from its own basic corporate mission; that "joint venture" is now on a 100-0 basis.

In the remaining two cases the initial agreement provided that either of the two parent firms might purchase up to 50 percent of the other's interest based on assets, not earnings, at three-, five-, and seven-year intervals. In these two joint ventures, both parents have benefited, the technology contributor achieving a shorter-term investment recovery and the marketing contributor achieving a longer-term market interest with better technology and a better product.

Where such an option is not specifically spelled out in the initial joint-venture agreement, a series of power struggles and conflicts is likely. This option was agreed upon at the outset in one joint venture when an aerospace firm provided the advanced welding technology for the other partner's existing worldwide welding products business. The welding products company bought out the aerospace firm's interest after three years. The welding products firm wanted a long-term position in the welding business it had been engaged in for the past 30 years, while the aerospace company recovered its sunk investment and profit objective from this shorter-term joint venture.

Profit-Loss Distribution

Every agreement should have clauses addressing the areas of profit distribution and means for coping with failure. A parent with a strong financial position usually prefers to have profits or earnings plowed back into the joint venture while a financially weaker parent usually desires current dividends. In 4 of the 19 joint ventures, the agreement called for a percentage of the profits to be earmarked for the strong parent's more basic R &D efforts.

Every venture should also be structured to allow for the withdrawal of a partner. Five of the ventures studied had an "auction agreement" under which either partner might offer to buy out the other. The party receiving the offer has two options at that time: either to accept the offer or to buy out the offerer on the offerer's own terms. This type of agreement helps keep both partners honest and guards against efforts to cut losses.

To cover the event of failure or liquidation of the joint venture, many agreements provided that the technology and patents go back to the contributing parent and the fixed plant investment go to the other parent. This provision usually conforms with each parent's original contribution.

Joint-Venture-Agreement Summary

The considerations included in the joint-venture agreement must always be specifically tailored to the mission and objectives of the particular joint venture. But from a macro perspective, a number of overall guidelines or principles that will assure a fair deal to both parents should be considered. These include:

—A precise definition of the joint-venture mission.
—Initial and future contributions of each parent: technology and patents, manufacturing-plant equipment, capital and personnel.
—Initial equity position and clauses for changes over time.
—Performance clauses for each parent.
—Composition and policy decisions of the board of directors or the partnership committee.
—Distribution of profits or absorption of losses.

5
Operating Practices

The 19 field-study cases also yield some fruitful guideposts to running the fledgling joint venture. This last stage of the process will be replete with problems if the previous negotiation and agreement stages were not well conceived.

Board of Directors

Since every joint venture is an appendage to two independent corporations, there must obviously be a group of executives to represent each firm's interest in the joint venture. Among the joint ventures studied, this group was usually called a "board of directors," but it was not uncommon to find it being called a "policy committee" or a "partnership committee."

In every case, the number of board members each parent provided was in direct proportion to that firm's percentage of ownership. In a 50-50 joint venture, each parent usually contributed three or four members to the board. As a venture changed from a 50-50 to a 51-49, 60-40, or 80-20 ratio, the respective composition of the board would change accordingly.

Board members were often described by respondents in the study as identifying too strongly with their respective parent firms and not strongly enough with the new entity. The split-board danger is always prevalent in a 50-50 joint venture. Including outside board members lessens this danger and can also be a tie-breaking mechanism.

Where the joint-venture manager has a voting role on the board and is originally from one of the parents, he is perceived as representing that parent. One joint-venture charter provides that the joint-venture manager shall vote only in a deadlocked voting situation. In this case the manager is not from either parent. The mere presence of this mechanism is believed to work psychologically to prevent deadlocks.

To avoid voting solely along parent lines with decisions thus ending in deadlocks, 8 of the 19 joint ventures studied have two or three board members from outside the parent corporations. Many joint-venture managers believe these outside directors often bring a highly needed objective approach to the major issues faced by the board of directors. Four joint-venture managers commented that they desired a few outside members on the board but had difficulty finding persons mutually agreeable to both partners.

The desire by either parent firm to have an outside board member, or in essence an arbitrator, may signal some basic problems in the marriage. Arbitration may be inferior to rational discussions by the disputing parties who really know what is operationally needed and what the real issues are.

The board must be more involved in the joint venture in the beginning stages than when it is an established autonomous business. Any heavy involvement of the board in any of the stages should always be limited to concerns over policy. The board should never be involved with operational problems. Avoiding the assumption of this role may be difficult for some board members in the early stages, when operational problems are greater, but their involvement at that level would be counterproductive to the success of the joint venture.

The Joint-Venture Manager

The choice of a leader for any organization is a critical decision. The leadership for most *internal* ventures has come from within the sponsoring company. But a senior vice president at one of the joint-venture parent firms commented, "Neither one of the parents had a single person who both understood the technology and had previous hard-nosed business experience, but we wanted the best man possible. We found him outside our two parent firms; he was expensive, but definitely worth it."

The selection of a qualified joint-venture manager is often made more complex when initially each of the parents own 50 percent of the venture and picking a manager mutually agreeable to both parents is not a simple task. It is not surprising that 6 of the 19 joint ventures in our study resolved their conflicts by selecting managers from outside the two parent firms. An alternative solution to the problem occurred in one instance when the president of the joint venture came from one parent and the chairman was drawn from the other.

What are the right qualifications for a joint-venture manager? Clearly, previous responsibility for profit and loss is essential. In our study, all six of the joint-venture managers recruited from outside the parent firms had previous division-manager or small-business P&L experience. Seven of the other managers had previous P&L experience within one of the parents as either a division, department, or internal-venture manager. Therefore, 13 of the 19 venture managers (13 of the 17 successful ones) had previous experience managing a total business with P&L responsibility.

The two unsuccessful joint-ventures were headed by technical Ph.D.'s who had no previous total-business or P&L experience. A parent of one of these failures commented: "If we had had a hard-nosed businessman running the venture, who was more applied and not theoretical, it may have been a winner."

In these 19 case studies, 6 joint-venture managers had the title of president and the remaining 13 held the title of general manager. Two of the presidents claimed that title provided more "organizational clout" than the general manager title.

Previously conducted personal interviews with over 60 internal-venture managers[8] definitely indicated that nearly all the joint-venture managers in our study have comparatively more operational freedom to control or manage their new product or new business than the internal-venture managers possessed; it would therefore seem that a separate legal entity, separate financial capitalization, and often a separate geographic location added to the increased autonomy. Sixteen of the 19 joint-venture managers believed they had sufficient autonomy but 4 expressed discontent with having two audits (one by each parent) and with frequently needing to satisfy both parents.

Location of the Joint Venture

As with some newlyweds, establishing a separate home away from either parent may be a slow process for a joint venture. In the 19 cases under discussion, since the roles and skills supplied by each parent were not solidified until after the honeymoon period, the joint venture was usually initially located on the premises of one or both parents. Not until each firm was thoroughly acquainted with the other, and their roles had crystallized, did joint-venture activities come under one roof. Since one of the parents usually provides technology and the other marketing skills, a geographic separation of these basic activities is inconvenient, but the early start-up activities can proceed.

Even when the technical and commercial personnel were housed in the same geographic location, the specific choice of location was determined by a number of factors. The need to be near the technology-contributing partner was the reason given by 12 of the 19 joint ventures that chose to conduct their activities at or near the technical partner's home location.

The frequent need for the technical interface was also exemplified by one joint venture that located itself near a major university that had the most extensive research program in the selected area. The availability of unoccupied or excess space at one of the parent's locations was cited by four of the joint ventures as the reason for their current locations.

Three of the joint ventures are still performing the technical, engineering, and production tasks at one parent firm and shipping the finished product to the other parent firm for marketing. The managers of these three joint ventures did not envision any independent third location in the near future, doubtless because the technical, production, and marketing factors in their ventures were easy and natural "fits" into the existing production and marketing systems of the parent firms. These parent firms moreover are geographically within one to two hours driving time from each other.

[8]James D. Hlavacek, "Toward More Successful Venture Management," *Journal of Marketing*, October 1974, pp. 56-60; Richard M. Hill and James D. Hlavacek, "The Venture Team: A New Concept in Marketing Organization," *Journal of Marketing*, July 1972, pp. 44-50; and James D. Hlavacek and Victor A. Thompson, "Bureaucracy and New Product Innovation," *Academy of Management Journal*, September 1973, pp. 361-372.

Selecting and Integrating Key Personnel

Every joint venture's success is of course highly dependent on the people responsible for its actual activities. Generally the key personnel were relatively young (25-40), aggressive, and judged to be superior in competence. Strong top management involvement in the joint venture from the beginning may have made it easier to obtain the best qualified people from each of the parent firms. Every manager who responded to our question believed his key employees had more freedom and larger challenges than they would have had if the venture were internal to one of the two partners.

Eight of the joint ventures studied had difficulty convincing the key technical man from one of the partners to leave with "his" technology and join the new organization. A number of the key technical people returned "back home" after being exposed to the more rapid activities at the joint venture. The technical inventor still wanted to invent and was uninterested in the cost-profit orientation of the engineering, product-development, and market-development functions. In a number of these cases, this problem was partially solved by using the technical inventor as a part-time technical consultant.

The key technical man is often as critical to the new venture as the technology he created is. Therefore the problem of technology transfer is partially one of transferring key technical people to operating units closer to the marketplace.

In 14 of the 19 joint ventures studied, *all* key personnel except the joint-venture managers came from the two parent firms. Six of the key production managers came from outside the parent firms, one, an earlier licensee of the technology, coming all the way from Europe.

One explanation for this phenomenon may be the strong emphasis the two partners placed on the technology and marketing contributions to be made to the joint venture. Production, a buffered area between the technology and marketing areas, may have seemed easier to manage, but when production hurdles blocked the way and it became known that remaining within the bounds of the earlier projected costs would be difficult, it became obvious that production expertise was very much needed.

If a large number of the key functional people must be recruited from outside either partner firm, this is an early warning sign that a mismatch of skills between partners has occurred, since a major reason for the joint venture is to utilize each partner's key skills.

During the early start-up period, venture managers frequently "purchased" part-time professional assistance from the parents—and were charged for it. This time period and part-time assistance also provided the prudent venture manager with an opportunity to observe who the better-qualified key personnel might be—especially unfamiliar people contributed by the partner firm the venture manager did not come from. Conversely, the part-time exposure also allowed the part-time employee to explore the more entrepreneurial environment of the joint venture.

Such part-time assistance, usually charged to the P&L function on a man-hour basis, is especially beneficial to a joint venture formed by a large and a small firm. The small firm often does not have the in-house legal and patent counsel, advertising personnel, technical support, and possibly lobbying expertise that are typically more essential in the premarital period.

The use of a nuptial vocabulary here is not accidental. Every joint venture is a marriage, and its premarital visions of blissful harmony may be dispelled after the wedding. Fourteen of the 19 joint-venture managers believed that the technical people integrated with their functional counterparts more rapidly than members of any of the other functional groups did. The remaining five managers felt that all personnel integrated at about the same rate. The commercial-business personnel were slower to integrate than the technical people because of differences in financial and policy procedures in their parent organizations.

Eleven venture managers made the general comment that "the younger in mind integrated more rapidly" and wanted to be evaluated on actual performance, not by seniority or job classification. Respondents almost unanimously agreed that the technical personnel and younger men identified much more strongly with the joint venture than with their parent firm. Venture managers and key production personnel who were hired from outside the parent firms also identified strongly with the joint venture.

Not surprisingly, when the joint venture was located geographically on the premises of one parent or near it, the personnel from that parent identified with it rather than with the joint venture, since their past professional and social relations were less disturbed by their changed circumstances. When the affairs of the new business went unfavorably, the desire to return "home" occurred much more frequently when the parent was close by.

The homesickness factor was experienced by fewer key personnel when the joint venture was at an independent location a good distance away from either parent. In two of the joint ventures studied, when a technical man defected to his parent firm, he was not accepted and went elsewhere.

The problems of homesickness and identity were particularly severe during the preformation period of development when the personnel assigned to the joint venture were still on the payroll of their respective parent firms. The pain was reduced when the joint venture was legally formed with its own capitalization, budget, and payroll.

When the legally and financially separate joint venture was geographically located near or with one of the parent firms, the employees at the parent firm perceived the joint-venture personnel as quite separate from their firm. One venture manager said, "We are received as bastard children are at the family reunion—they speak to us superficially and tolerate us because we may succeed."

Performance Evaluation and Compensation

One would hypothesize that when two firms contribute personnel and equity to a new third firm, performance evaluation and compensation would create a number of problems. Quite the contrary was found to be true: none of the joint-venture managers in our study thought these considerations had presented any previous or current major problems. The parent firm that supplied the greater number of people and/or the physical location also had its evaluation and compensation system adopted for personnel *below* the president or general manager level.

Four presidents said it was "lucky" that both parents used the same rating systems. Many felt that differences in accounting methods and fiscal year dates were more important areas of concern and more urgently required agreement than the actual performance review and compensation method.

The joint-venture manager usually evaluates key employees, using their respective jobs at the parent firms as general guidelines. He then discusses his assessments with the senior management at each parent firm before arriving at a final decision. The manager himself is evaluated against the common criteria of return on investment (ROI) profitability, and his budgeted performance plan.

Only 5 of the 19 joint-venture managers had an equity position in the joint venture, whereas the remaining 14 had the more traditional stock option performance plans offered by the parent firms. The five venture managers with an equity position are all heads of new entities created by relatively small parent firms.

None of the personnel below the level of the manager had an equity position in any of the joint ventures studied. One of the joint-venture managers with an equity position said, "At times I feel guilty when others are working 60 to 70 hours per week for a salary." Another manager with an equity position claimed he was attempting to obtain an equity position for the key personnel and a stock plan for all employees in the joint venture.

6
Antitrust Considerations

Possible violations of the antitrust provisions of the Sherman and Clayton acts should be seriously considered by a large firm about to enter into joint ventures involving new-product or new-business development. The Justice Department had earlier compared joint ventures with mergers but has since recognized a number of distinct differences between these entities and their impact on competition.

> A horizontal merger *reduces* the number of competitors by one or more while vertical mergers and conglomerates tend to leave the number unchanged; a joint venture *increases* the number of competitors.[9]
>
> A merger often *reduces* the number of customers available to suppliers while a joint venture tends to *increase* the number of customers.[10]
>
> In a merger, two or more companies combine *all* of their assets to create a new entity. In a joint venture, two or more companies combine *less* than all of their assets to create a new entity. . . . The assets combined may be either capital, technical skill, patents, marketing ability, or management experience, or a combination of some or all of them.[11]

Thus joint ventures for new-product innovation are not open to one of the major criticisms of the corporate merger: "functionless and dangerous growth." Mergers combine existing entities while joint ventures create more productive new entities. The joint venture also tends, not to eliminate market barriers, but rather to facilitate the aggregation of sufficient managerial and financial resources to overcome them.

In arguing for the legitimacy of joint ventures, corporations have de-

[9]Jules Backman, "Joint Ventures in the Light of Recent Anti-trust Developments: Joint Ventures in the Chemical Industry," *The Antitrust Bulletin,* January-April 1965, p. 9.

[10]*Ibid.*

[11]Lewis Bunstein, "Anti-Competitive Joint Ventures," *The Antitrust Bulletin,* January-April 1965, p. 25.

emphasized the financial advantages derived through collaboration and have asserted the technological necessity of their combinations.[12] But there is also a strong argument to justify joint ventures on the grounds of financial risk. Because of increasingly large technical and marketing costs required for developing and nurturing a new product, the financial risk of new-product and market development is, as noted, becoming too large for any single firm to undertake.

Yet, in the investment banking field, a pooling of financial risk is considered common and perfectly lawful even though the investment banking firms cease to be competitors when they form "syndicates."[13] Such syndicates are formed because no one bank commands sufficient financial resources to carry the risks of marketing a block of securities. Therefore the initiator or "lead firm" calls in a host of rival firms to share the risk that the decline in market values will be larger than the small "spread" between the offering price to the public and the proceeds to the issuer.

In this study, every joint venture between two large firms resulted in one partner contributing largely technical capabilities and the other commercial or marketing expertise. The two firms were combining, not similar assets, but dissimilar ones to create a synergistic new entity.

Joint-venture cooperation between a large and a small firm is favorably viewed by the Justice Department. Most cooperative efforts between two small firms are exempted from the antitrust laws by the Small Business Act of 1964.[14] A small business firm is defined as one "which is independently owned and operated and which is not dominant in its field of operation.[15]

One joint-venture manager said, "We are not sure what the antitrust implications of our venture may be. But if the venture goes well. . . we are prepared to dissolve it and the parents will pursue the new business as separate competitors." When two parties propose a particular joint-venture activity that they fear may come under antitrust scrutiny, the Department of Justice has available its Business Review Procedure whereby the proposal may be assessed in advance for possible antitrust difficulties.[16]

If one of two large parent firms in a joint venture later becomes engaged in producing and/or marketing a product identical to that of the joint venture, the Justice Department may dissolve the joint venture. The Justice Department looks favorably toward a joint venture between two large firms when one of the firms previously attempted but failed to profitably develop the new business or new product alone.

[12]Paul Dixon, Chairman of the F.T.C., "Joint Ventures: What Is Their Impact on Competition?" Address before the Economic Club of Detroit, March 12, 1964; *The Antitrust Bulletin,* March 1964.

[13]*United States v. Morgan,* 118 F. Supp. 621, 690 (S.D.N.Y. 1953).

[14]15 U.S.C. § 638 (d) (3) 1964.

[15]15 U.S.C. § 632, 1964.

[16]Statement of Donald I. Baker, Director of Policy Planning, Antitrust Division, Department of Justice, before the House Committee on Science and Astronautics, Subcommittee on Science, Research, and Development, on Science, Technology, and the Economy, April 18, 1972.

Four of the joint ventures in this study were the result of unsuccessful solo attempts. Underlying domestic new-product joint ventures is the belief that an alliance between two firms will be more efficient in terms of both significant product innovation per dollar spent and the time in which the necessary product innovation can be fully commercialized.[17]

[17]Special Conference of Invited Experts on Air and Water Depollution: Role of Industrial Property, Innovation and Competition, #112 Research Institute of George Washington University, March 31, 1970.

7 Some General Guidelines

In addition to the specific practices enumerated above, some overriding principles also emerge from our study. These can perhaps be stated most simply as a series of "do's" and "don'ts." Failure to follow them is likely to lead to disturbed (and disturbing) children and to divorce.

1. Be completely candid with your future corporate "spouse" with respect to the strengths and weaknesses you would bring to the marriage, the state of development of your technology, the personnel who would and would not be available, and so on.

However, we found that it is a good idea to introduce total candor gradually, especially when a large and a small firm are exploring possibilities for an alliance. Too much honesty, too early, may prove mischievous. As the executive vice president of one large responding company noted, "Our 'openness orientation' was first interpreted as a signal that we wanted to impose our methods on the joint venture."

Companies exploring the possibilities of a joint venture for the first time, we found, tend to be more secretive and suspicious. Small firms with the desired new technology are also inclined to "hold their cards close to the vest," since exploiting that technology is of the essence to them. Interestingly, when they do disclose their technology eventually, their larger prospective partners are sometimes surprised by its superiority over their own technology.

2. Beware of differences in corporate goals. These differences are likely to be most severe between large and small firms. When a small and a large firm co-venture, the new enterprise may represent 50 to 80 percent of the small firm's growth objectives and only 5 to 10 percent of the large firm's future growth expectations. One executive with a large firm said: "Our key people are spending 30 percent of their time on a new business that will probably never contribute more than 5 percent to our corporate sales and earnings."

3. Be certain that you will get something you really need. In 14 of 19 cases studied, the managers asserted that each partner must feel he would receive an otherwise unobtainable competence.

4. Avoid duplication of assets; it is wasteful and leads to conflict, especially if both firms are strong in one area. It is a good idea to have an independent outside appraisal made of the technological, production, and marketing assets that are proposed for the new venture. This will both expose

duplications and provide further, objective assurance of the validity of the prospective match.

5. While we can all use more money, one thing a firm should not seek in a joint venture is *only* money. Among the cases we investigated, two were of ventures in which one firm provided only capital; one of these ventures is in limbo and the other has been terminated. In two other instances in which the parent firms obtained capital from a third party, both joint ventures failed.

6. Have a long engagement period. Lengthy discussions early in the negotiations are especially rewarding. The probability of a successful venture increases in proportion to the understanding each partner acquires of the other. One firm with two previous failures under its belt has learned the hard way that it is useful to achieve thorough familiarity; the firm now exposes representatives of the prospective partner to its facilities, product lines, working conditions, operational procedures, and the relationships between the firm's various management echelons.

7. As part of being prudent, it is wise beforehand to consider alternatives to a joint venture, such as internal development, licensing, or acquisition.

8. Be realistic about the time frame. This injunction is directed especially to smaller firms that can move fast and may be prone to impatience. As noted, as long as three years may elapse before a new venture moves from fantasy into a reality.

Overall, the twin watchwords are *caution* and thorough *investigation*. Those who fail to obey them will surely repent afterward.

8
Conclusions

With such negative economic forces as slower GNP growth, greater capital formation problems, and continuing energy and materials shortages, the risk level for product innovation is mounting. Corporate growth, stability, and even ultimate survival become more difficult when the interrelationships of these factors create more unstable conditions than many managers have ever encountered. New-product development and market growth often slacken in such an environment at a time when both domestic and foreign firms are intensifying their competition.

Given this perspective, Peter Drucker predicts that the flexible joint-venture approach will become increasingly important as it becomes better understood.[18] The lack of attractive acquisition candidates, the desire for less long-term debt to achieve more liquidity, and an emphasis on cash flows are making the joint-venture approach an increasingly desirable means toward product innovation. More managers are seeing this pooling of technology, production capabilities and know-how, marketing skills, capital, and managerial expertise as a synergistic and less risky approach to product innovation.

Not that joint ventures are a cure-all for a languishing product- and market-development function. Rather, they represent a new organizational approach that opens avenues increasingly being barred to individual firms. In serving such a role they may enable new technology to be utilized, improve productivity, and bring about a more efficient deployment of increasingly scarce corporate resources.

[18]Peter F. Drucker, *Management: Tasks, Responsibilities and Practices* (New York: Harper & Row, 1974), p. 720.